FORCES, MOTI
PRESSUI

An Introduction to Key Stage 3 Physics

Written and Illustrated by W Booth
Edited by Max Edward

ISBN-13:978-1546637547

Printed in dyslexia-friendly typeface, and written with an informal style, the concepts of forces, motion and pressure are uncovered in a way appealing to any budding Physicist thirsty for knowledge.

Practical experiments using household items help cement the basics of Key Stage 3 Physical Science and the specially designed layout makes revision easy.

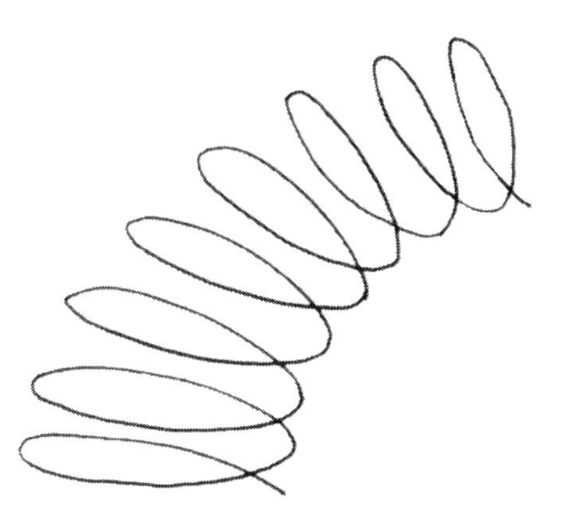

CONTENTS

SPEED

How slowly does a snail crawl? How fast can an Olympic Athlete run? What speed does a space shuttle travel? How quickly can you read this book?

MEASURING SPEED

To answer any of these questions, we first need to identify a few key points.

Let's start with the snail. How can we work out the **speed** it moves? What information do you think will help us calculate this?

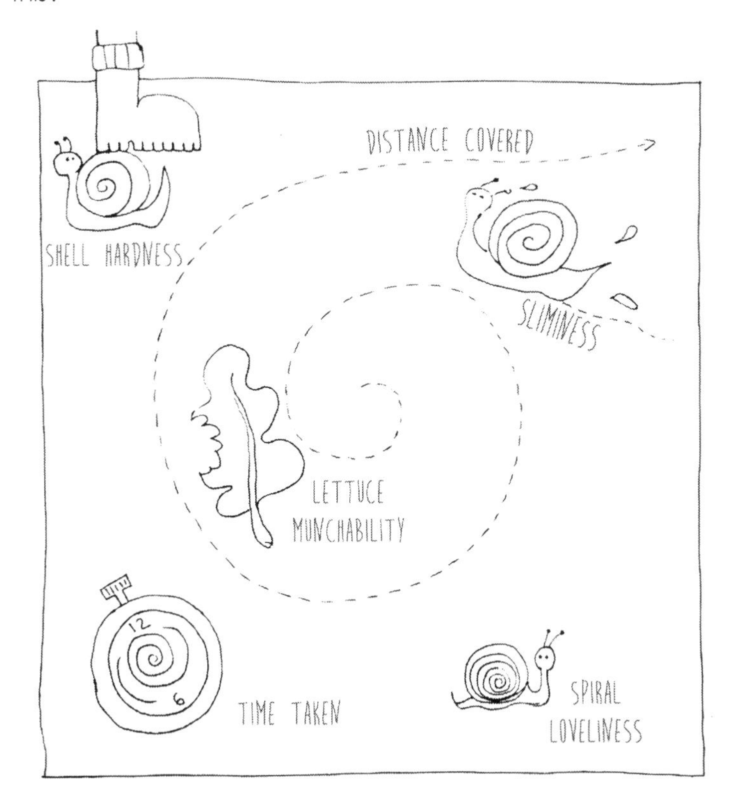

In fact, to calculate the speed a snail crawls, we only need to know two things: **time** taken and **distance** covered.

How far did the snail crawl?

and

How long did it take?

We can measure the speed of *anything*, if we know the time it took to cover a particular distance.

So, what units do we use to measure speed? You may have heard of speed being measured in miles per hour (mph), or kilometres per hour (kph), but in Science we tend to refer to speed in **metres per second** (m/s).

Distance is measured in *metres* and time is measured in *seconds*.

Let's say our snail is a turbo-powered snail and he covered the distance of 3 metres in one second.

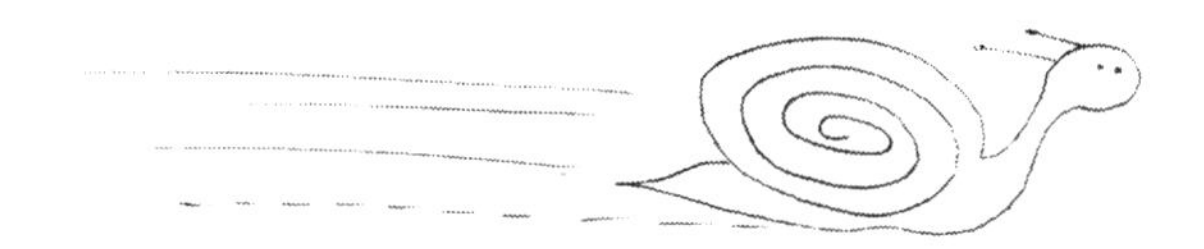

The formula would look something like this:

Speed of snail = 3 metres per second

We can replace the 'per' with a 'divide' sign as it means the same thing:

Speed of snail = 3 metres ÷ 1 second

So, now we can see how speed is calculated:

SPEED = DISTANCE ÷ TIME

Let's try this with some other measurements:

An Olympic Athlete broke all records and ran 100 metres in just 5 seconds! But, how do we measure her speed in metres per second?

As the question asks for speed, we can easily slot our two known values into the formula:

SPEED = DISTANCE ÷ TIME

SPEED = 100 metres ÷ 5 seconds

And 100 ÷ 5 = 20, so...

SPEED = 20 m/s

The distance covered over a specified time is actually **average speed** (as it may vary over the time but we've no way of knowing). However, we generally refer to 'average speed' as just 'speed'.

So, how about a space shuttle?

NASA says that in order to stay in orbit, a space shuttle must maintain a speed of 28,000 km per hour. If we want to compare this to our snail and Olympic Athlete (in Science we love comparing!), we need to convert it to the same units, ie, metres per second.

First, we need to convert the distance part of the value to metres. So, 28,000 km into metres? We just need to multiply it by 1000.

28,000 km – 28,000,000 metres

Then, to express an hour as seconds, we multiply it by 3600. There are 60 minutes in an hour, and 60 seconds in a minute. 60 x 60 = 3600.

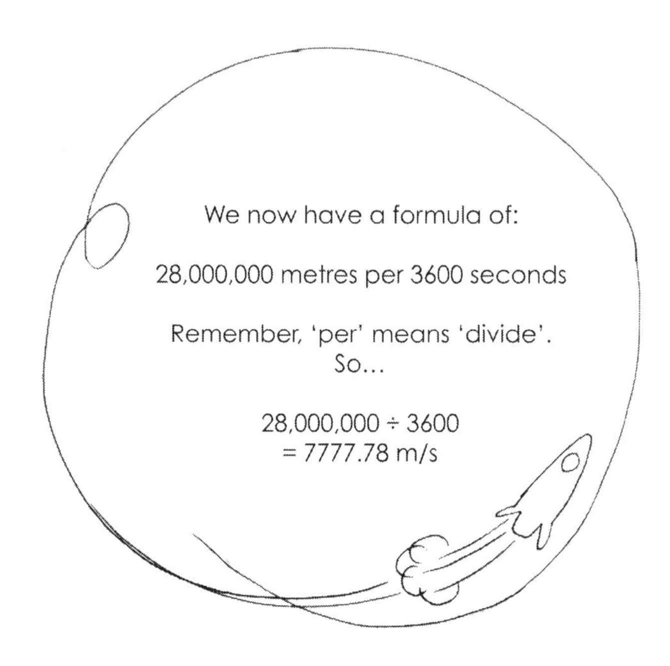

A little bit faster than our snail and Olympic Athlete!

So, how about how quickly you can read this book? Well, if you've started a timer already then you can calculate that for yourself (in words per minute) once you've finished all 4373 words!

Try this with a friend

One person hold a tennis ball about a metre off the ground. The other person hold a piece of paper at the same height.

At the same moment, both drop your object.

Which took the shortest amount of time to hit the ground? Which travelled fastest?

Less TIME means higher SPEED

SPEED-TIME GRAPHS

Speed of a journey over a period of time can be illustrated using a graph.

A **speed-time graph** shows how the speed of an object *changes* over time.

When the **gradient** of the line is level, this indicates that **velocity** (speed in a particular direction) remains constant.

SPEED and **VELOCITY** are words with a similar meaning.

SPEED: The rate at which an object covers a distance.

VELOCITY: The rate at which an object covers a distance *in a particular direction*.

If the gradient rises steeply, this means the velocity is increasing, meaning the object is **accelerating**. As the line is straight, this acceleration is constant (steady). And when the gradient falls the object decelerates.

If velocity is measured in m/s, then how do we measure acceleration?

Acceleration is how the velocity changes over time. So, looking at the graph we can see it is measured in **metres per second** (y axis) **per second** (x axis).

Remember, 'per' means the same as 'divide', so...

Metres per second per second

= m/s/s

= **m/s²**

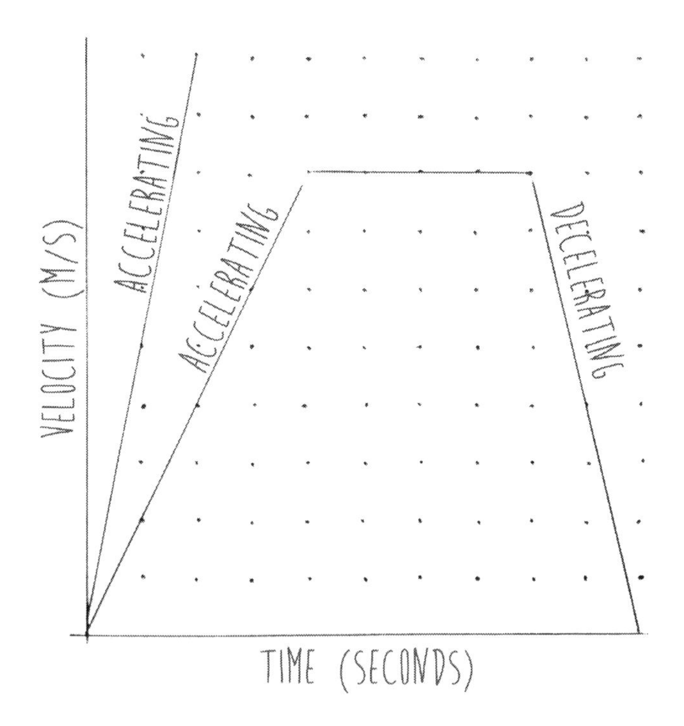

So, units of acceleration are m/s². Acceleration is a gradient on a velocity-time graph and can be drawn as a line.

-2 m/s^2 is negative acceleration which means the object is slowing down. This is also referred to as **deceleration**.

WORKING OUT DISTANCE

The graph gives us an easy way to calculate acceleration, but using this graph we can also calculate distance travelled.

Remember our original equation?

SPEED = DISTANCE ÷ TIME

An easy way to remember this is in a triangle:

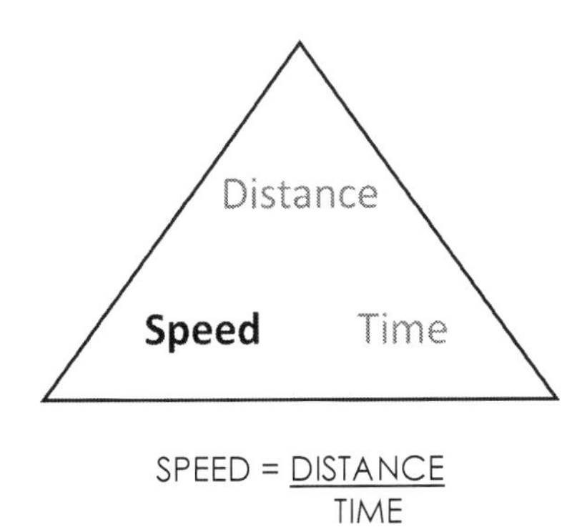

$$\text{SPEED} = \frac{\text{DISTANCE}}{\text{TIME}}$$

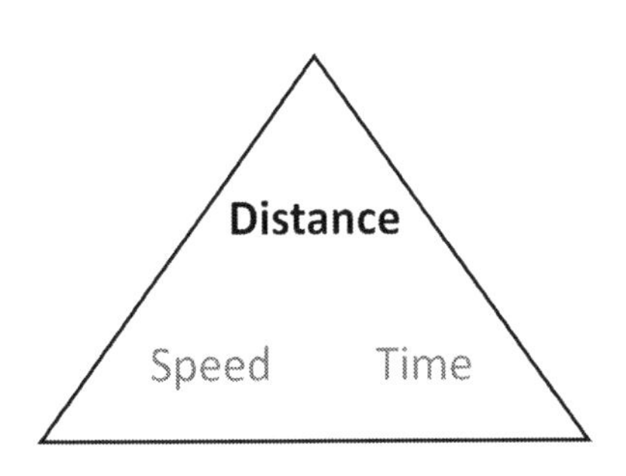

DISTANCE = SPEED x TIME

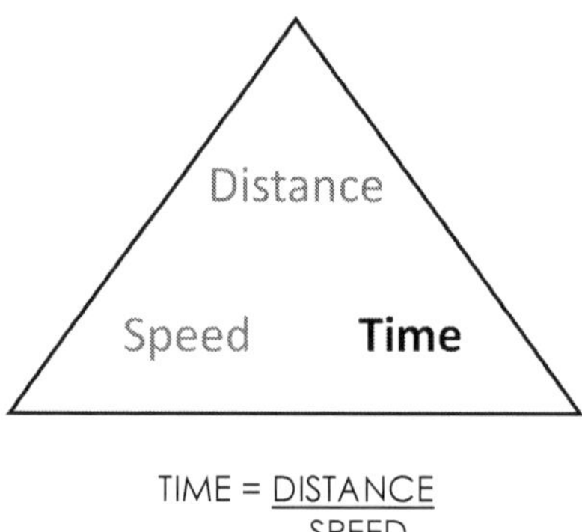

$$TIME = \frac{DISTANCE}{SPEED}$$

So, when the speed-time graph shows a constant speed (ie, a level line) then the distance covered is:

DISTANCE = SPEED x TIME

(LEVEL GRADIENT)

This is the same as the area under the line (see graph on next page).

But, what if the line is a sloping gradient? The gradient slopes when we are accelerating or decelerating, and again, we can calculate the distance covered because it is the area under the line.

As we are now calculating the area of a triangle, rather than the area of a rectangle, the formula changes to:

$$DISTANCE = \frac{SPEED \times TIME}{2}$$

(SLOPING GRADIENT)

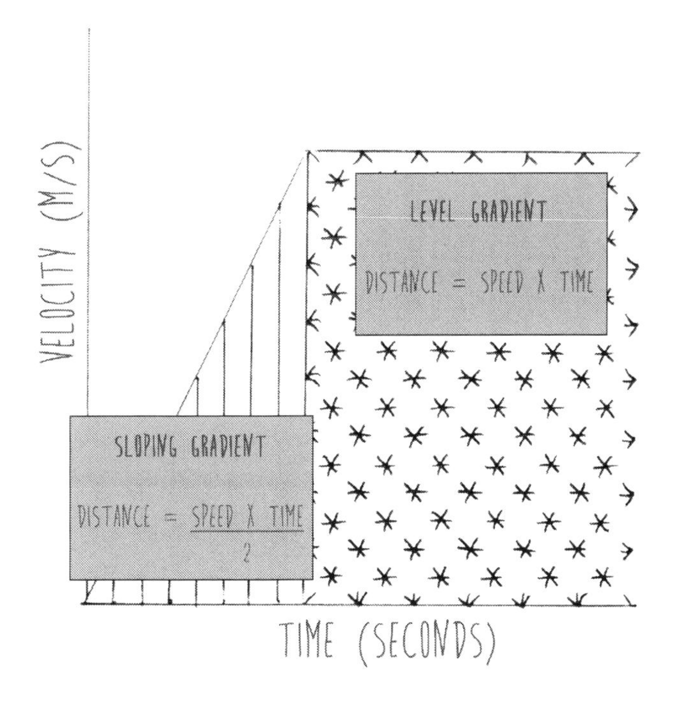

RELATIVE MOTION

Have you ever travelled in a train alongside another train of similar speed? It feels as though the train is moving very slowly, even though both trains are moving fairly quickly. This is because of their **relative motion** to each other.

Relative speed is something we can calculate quite easily.

If both objects are travelling in the same direction:

Relative speed =

speed of faster object – speed of slower object

If objects are travelling in the opposite direction:

Relative speed =

speed of one object + speed of other object

It doesn't matter whether the objects are travelling towards or away from each other, the calculation is the same.

FORCES

With a ping pong ball on a tray, try to make the ball move using as many different methods as you can. Make a list of all your ideas.

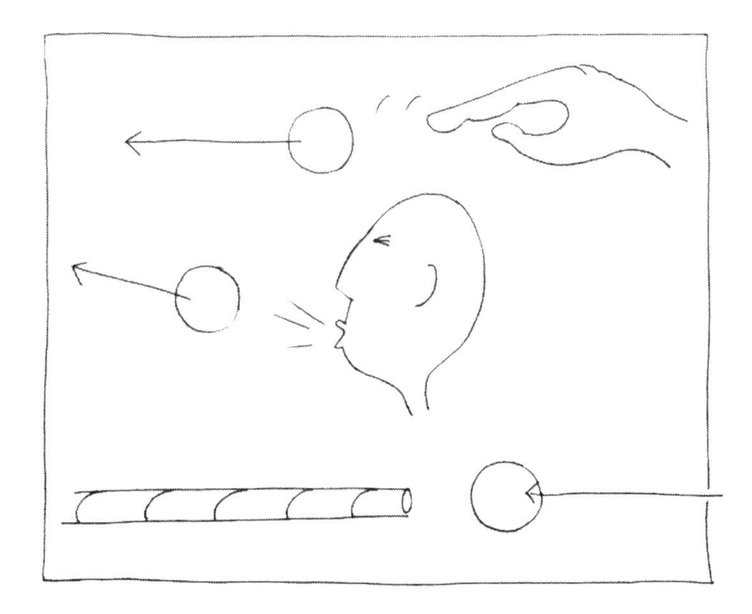

I'm sure you can think of many more than this! But in each case we are applying a **force** to the ball. Sucking with a straw would be called a **pulling** force and the blowing and flicking examples would be called **pushing**.

A force is a 'push' or a 'pull' between two objects.

The *pulling* force (for example in a cable, spring or rope) is called **TENSION**.

This may cause the cable to extend (stretch) or break apart.

All these forces on the ball are caused by **friction**. Friction is a force caused by the interaction of surfaces moving over one another. In the case of the blowing and sucking examples, we call this **drag** because one of the objects is a fluid (in these examples, air).

Did you notice the arrows in the previous illustrations? This is how we demonstrate force in a **Force Diagram**.

The arrow shows the **size** of the force (the longer the arrow, the larger the force) and the **direction** of the force. The arrow is usually labelled with the name of the force and its size in **Newtons** (N).

Can you match each variable to the unit used to measure it? If you get stuck, have a quick look through the previous pages – you will find all the units in bold.

Force	metres (m)
Distance	seconds (s)
Speed	metres per second (m/s)
Time	metres/second/second (m/s^2)
Acceleration	Newtons (N)

BALANCED FORCES

When two forces acting on an object are *equal* and *opposite*, we call these **balanced forces**.

If the forces acting on an object are balanced then it continues to move at the same speed and in the same direction. This means that an object which isn't moving will remain still.

However, don't forget an object can be moving even with no forces acting upon it.

Here are some examples of balanced forces:

- Boat floating on water
- Book on a table
- Snail sleeping on a wall

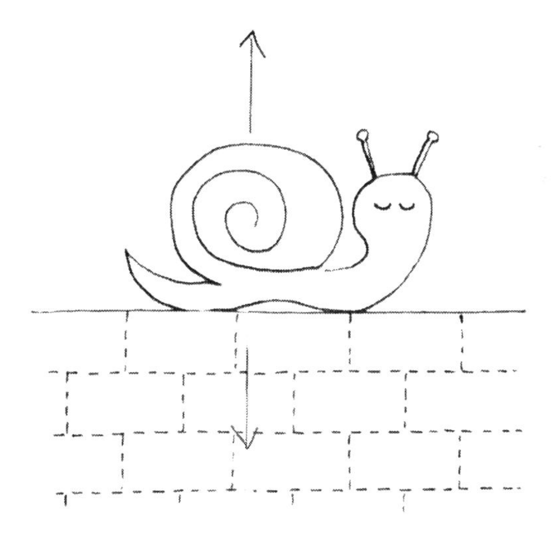

Can you think of any more?

When two opposing forces are NOT equal in size, we call them **UNBALANCED FORCES.**

If unbalanced forces act on an object then it will either:

- Begin to move, if it wasn't already
- Change speed and/or direction

For instance, a sudden gust of wind can cause a truck to speed up or slow down, depending on the direction of the force.

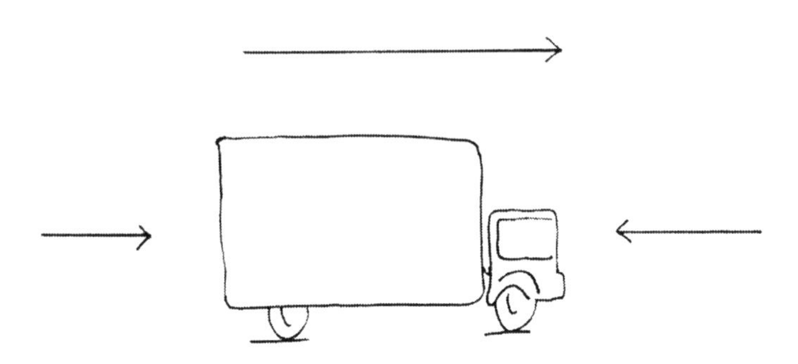

If the new force is from the same direction as the original force (eg, wind blows the truck in its direction of travel), then we add together both forces to find the **resultant force**.

RESULTANT FORCE is a single force which could replace all forces acting on an object and have the same effect.

However, if the new force is pushing the truck from the opposite direction (eg, wind is blowing against the direction of travel) then the resultant force is one force deducted from the other.

MOMENTS

When you are on a seesaw in the park, you use your feet to push up and then the force that pushes you down is gravity.

But, if you look closer, you will see that this movement is not just up and down, but you are following the arc of the see saw around a **pivot** (where the seesaw is attached to the ground). This *turning* effect of the force is called a **moment**.

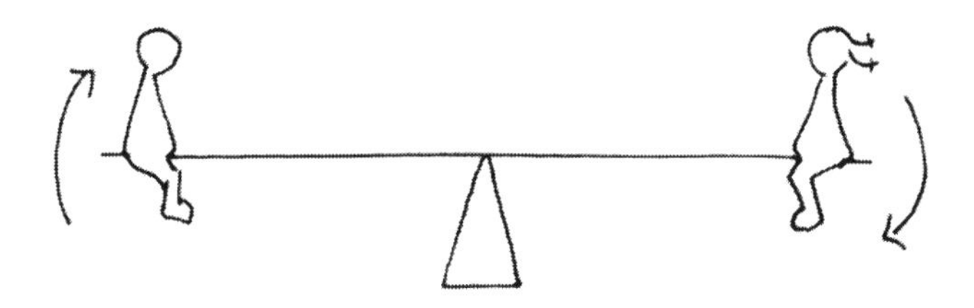

If you sit on a seesaw with a friend of just the same weight (mass) then if you are the same distance from the pivot, you should be able to balance the seesaw so the beam is level. This is because the *distance* from the pivot and the *force* each person exerts are both the same.

When two opposing forces are balanced, we call this **EQUILIBRIUM**

When the force is higher (for example, your dad gets on the seesaw) then you have probably noticed that in order for the seesaw to balance, the heavier weight must move closer to the pivot.

We can calculate *moment* by multiplying distance and force. So we can always calculate the exact values needed to balance a beam.

MOMENT = FORCE x DISTANCE

As we are multiplying force (measured in Newtons) and distance (measured in metres), the units for measuring *moment* are called Newton metres (Nm).

The moments on each side of a balanced seesaw are equal, but they are also opposite because they exert force in opposite directions – one clockwise and one anticlockwise. This is why they balance.

A seesaw is an example of a **simple machine**. We often use simple machines to allow something heavy to be moved easily because we apply a force from a greater distance from the pivot.

Can you think of examples of *simple machines*?

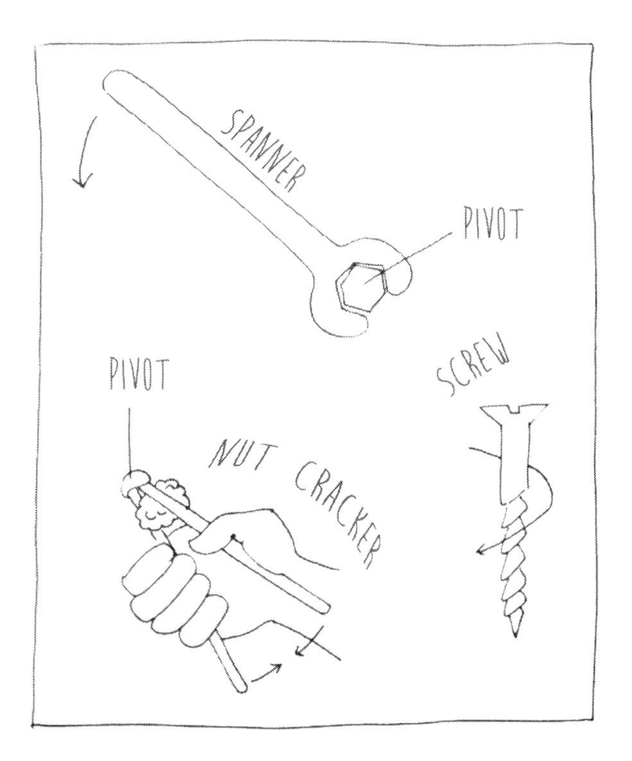

Simple Machines

DEFORMATION

As well as moving an object, forces may also *change the shape* of an object. A spring, for example, may be *squashed* (compressed) or *stretched* (extended). We call these stretchy objects **elastic**. An elastic object stores potential energy.

Other examples of elastic objects include:

- Hair bobble
- Catapult
- Tennis ball

A change in shape like this is called **deformation**. The more force exerted, the more the deformation. However, if you stretch or squash an object too much, it may never return to its original shape and may even snap. Deformation is often described as *STRAIN*. When an object is stretched permanently out of shape or it snaps, it is said to have passed its **elastic limit**.

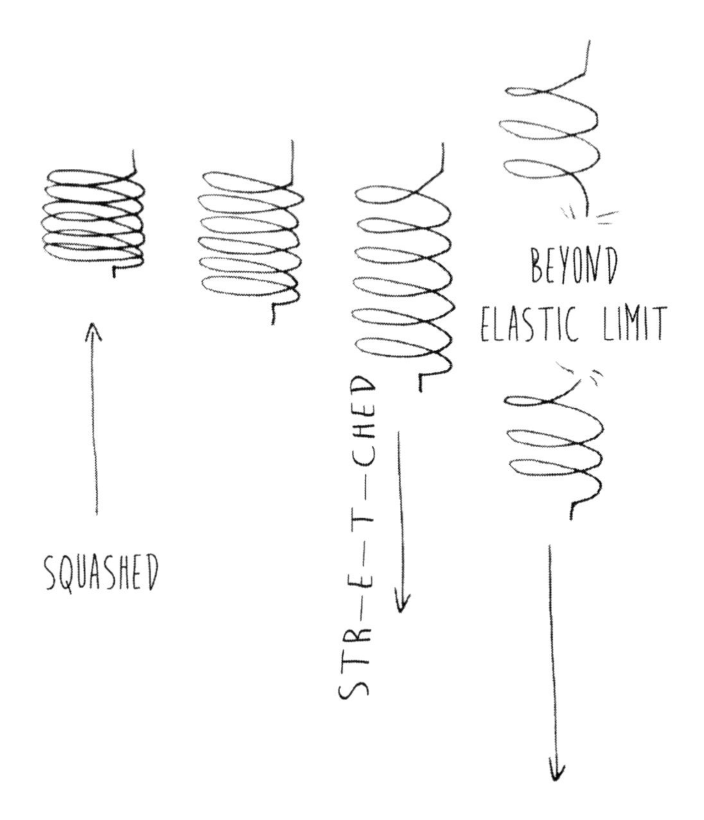

Deformation refers to any changes in the shape or size of an object due to either:

- an applied force – where energy is transferred through *work* done.

- a change in temperature – where energy is transferred as *heat*.

HOOKE'S LAW

The 17th century physicist Robert Hooke had a great interest in springs and everything elastic. He stretched and pulled them, experimented and plotted many graphs, so that we don't have to.

He found that there is a **linear relationship** between the force exerted on a spring (by hanging weights on it) and how much it stretches. A *linear relationship* means that when plotted on a graph, the line is straight.

The graph below shows a straight line that passes through the origin.

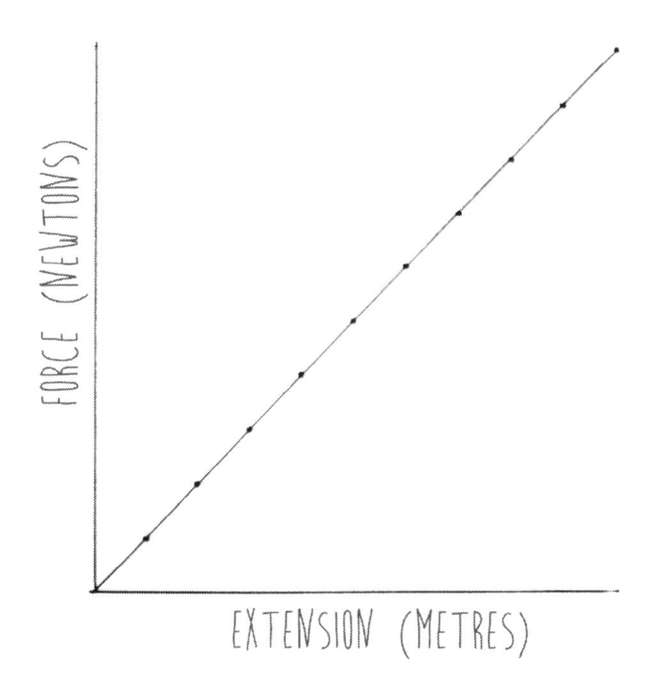

The steeper the line, the stiffer the spring.

Again we can make calculations based on the area under the line (remember doing this to work out distance covered?).

$$\text{WORK DONE} = \frac{\text{FORCE x EXTENSION}}{2}$$

Work is measured in Nm.

Hooke found this connection by experimenting and measuring his results *many* times. The fact that he could repeat his work and get the same results time after time, means his experiments were *repeatable*.

If you try this at school and are able to get the same results, then the experiment is *reproducible*. In Science we love repeatable, reproducible results as this indicates *proof* that our laws are as close to fact as possible.

However, this linear relationship is only true until the spring reaches its *elastic limit*. You know a spring has reached this limit if:

- Measurements no longer follow the predicted line of the graph
- The spring does not return to its original shape once weights are removed
- It snaps

Each material has its own elastic limit.

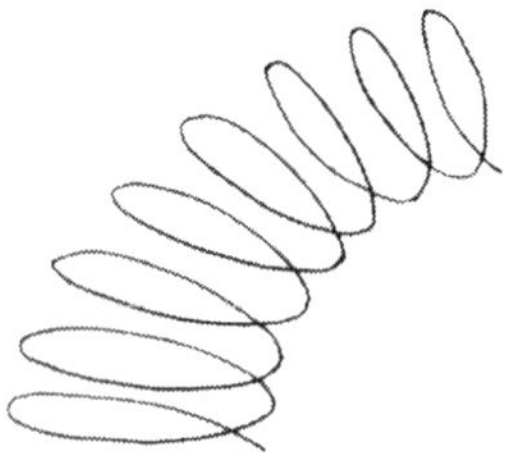

TYPES OF FORCE

When one object moves against another it causes **frictional forces**. Friction makes it more difficult for things to move. This can be helpful in several ways.

Helpful Frictional Forces

- Brakes against a wheel helping it to slow down
- Hands around a ball, allowing you to catch it
- Shoes on the floor stopping you skidding

Unhelpful Frictional Forces

A padlock is jammed closed. A lubricant is often helpful to help ease the friction.
Running through soft sand can be incredibly hard work

Try this

Push a toy car across the floor and let it go. Try this on different types of surface and make notes of how long the car will run once you've let go.

Good types of surface to try might be:

- Carpet
- Wooden floor
- Tarmac
- Grass
- A towel
- Table top

Which surface allows the car to run the furthest?

When surfaces are rough, friction increases. Smooth surfaces cause less friction.

A lot of friction between moving parts (in an un-oiled engine for instance) will cause *heat*. This effect will be familiar as the warmth you feel when you rub your hands together.

Air Resistance

The frictional force experienced when a moving object is opposed by air is called **air resistance**. The faster the object moves, the higher the air resistance.

As with friction, there are helpful and unhelpful types of air resistance. For instance, a parachute increases air resistance and so reduces a person's speed, allowing them to land gently.

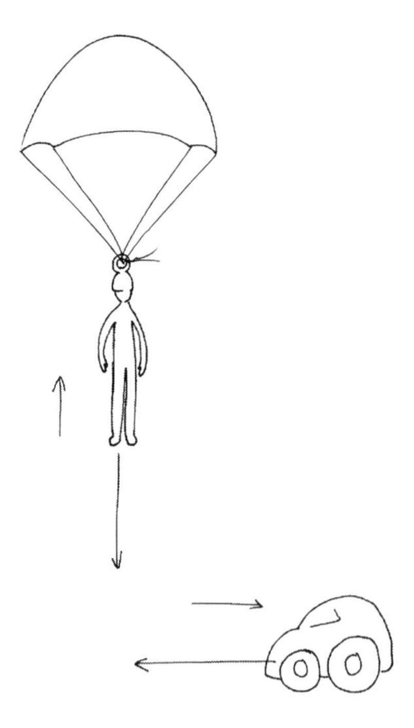

To reduce air resistance a cyclist may duck down so they offer a smaller surface area to the oncoming air. They may also wear a *streamlined* helmet to help them move faster.

See also how a car's shape is developed to reduce air resistance so it can travel more efficiently. Not only is this useful for increasing speed, but it means a vehicle can move more efficiently and so needs less fuel.

The forces discussed above involve direct contact (including by air). This type of force is called **Contact Force**.

A force which acts on an object without coming into physical contact with it is called a **Non-Contact Force.** For example magnetism and gravity.

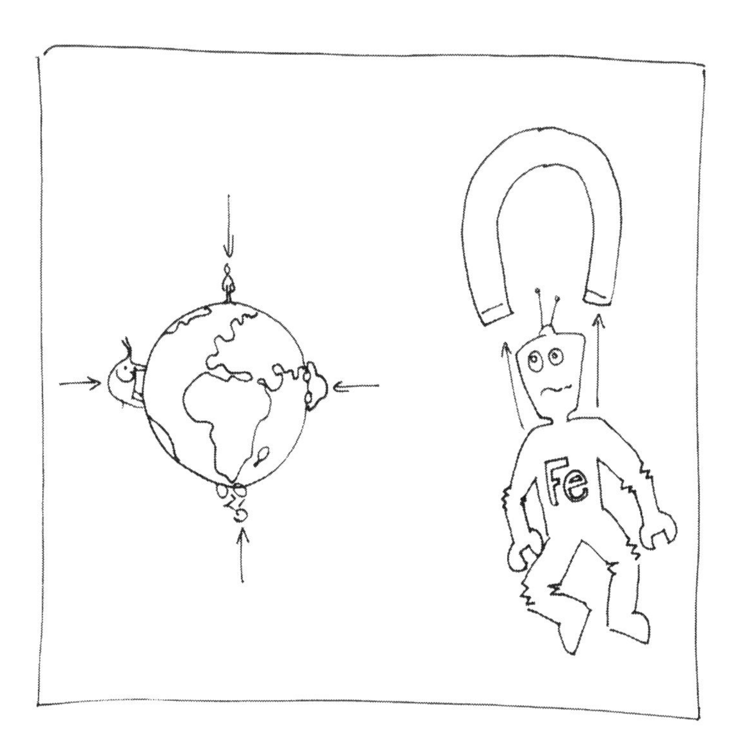

Gravity and Magnetism are known as Non-Contact Forces

WEIGHT, MASS AND GRAVITY

You will remember that when we measure forces, we use the unit *Newtons*. **Weight** is also measured in Newtons, because weight is a force (more on this later).

We are used to referring to *weight* in terms of kg or pounds, but in scientific terms, what we are actually referring to here is **mass**. Mass is measured in *kilograms* (kg) and describes the *amount of matter* an object contains. For example, a mountain contains more matter than a pebble, so it has greater mass. Whether this mountain is on Mars, Earth or the moon, it still has more mass than a pebble.

Objects with a large mass are called **massive**.

GRAVITATIONAL FORCE

When an apple falls from a tree, how many forces act upon it?

In the *force diagram* below, we see there are two forces acting on the falling apple. The air resistance opposing the object's approach towards Earth, but the largest force is the force which *pulls* it 'down' to Earth.

All objects have a force attracting other objects towards them. Even you!

This force is called **gravity**.

Gravitational force becomes more noticeable the more massive an object is, which is why objects fall to Earth quite readily, objects appear to float on the moon and why you are very unlikely to notice objects attracted towards your own body.

Gravitational pull will grow larger as an object increases in size. It is also affected by how close the objects are.

So, *gravitational force* increases when the objects are either:

- More massive, or
- Closer together

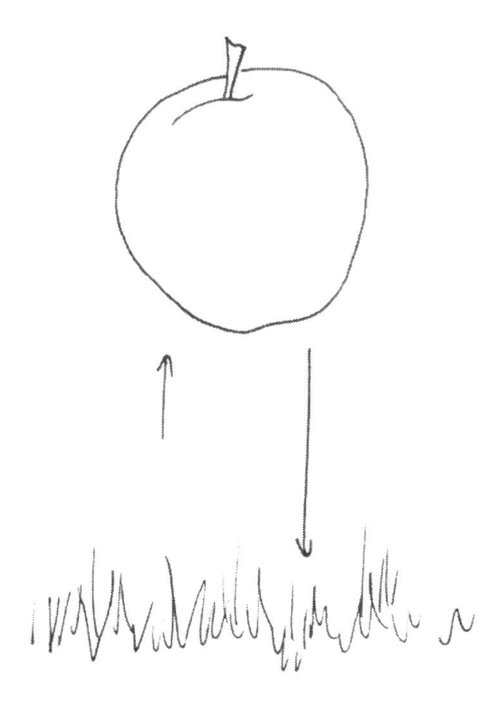

As mentioned earlier weight is a force. It is a force caused by gravity. On Earth, the weight of an object is the **gravitational force** between it and Earth. The more *massive* the object, the greater its weight.

Weight is the force of gravity on an object. It is measured in Newtons (N).

The mass of an object will remain the same wherever it is. But weight will change dependant on **gravitational field strength**.

Gravitational field strength is measure in Newtons per kilogram.

The area where other objects are affected by gravitational force is called the **field.**

We can see the relationship between weight, mass and gravitational field strength here:

WEIGHT = MASS x GRAVITATIONAL FIELD STRENGTH

Let's say a person has a mass of 70kg. Remember, this mass remains the same wherever she is.

We can approximate the Earth's gravitational field strength to 10N/kg, so to calculate the person's weight we multiply mass by gravitational field strength:

70kg x 10N/kg = 700N

The moon's gravitational field strength can be approximated to 1.6N/kg, so using the same method, we see the person's weight on the moon would be:

70kg x 1.6N/kg = 112N

Around a sixth of her Earth weight.

We can calculate weight on any planet (or massive object) given its gravitational field strength. Here are a few to try, just for fun!

Mars	3.8	
Neptune	13.8	
Uranus	8.7	
Saturn	10.4	
Sun	270	N/kg

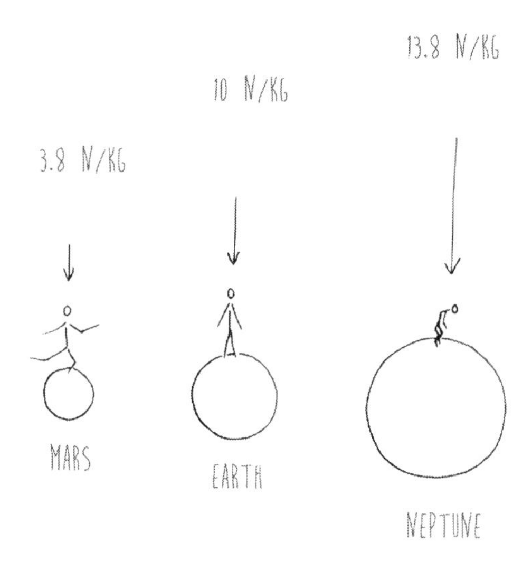

So, now when asked your weight, you have a choice of whether to give your Earth weight, your Mars weight or your weight on any planet you choose. Or more appropriately, you may wish to answer, "None of your business!"

PRESSURE

Which item would you choose to push into a wall to hold up a poster? A wellington boot or a drawing pin?

The reason a pointed object would make a hole so much easier is, assuming we are applying the same *force*, the **pressure** is so much greater.

With a wellington we are pushing on a much larger *area* and so the force is spread out and the pressure is *low*. With a drawing pin, the force is concentrated in a small area and so the pressure is *high*.

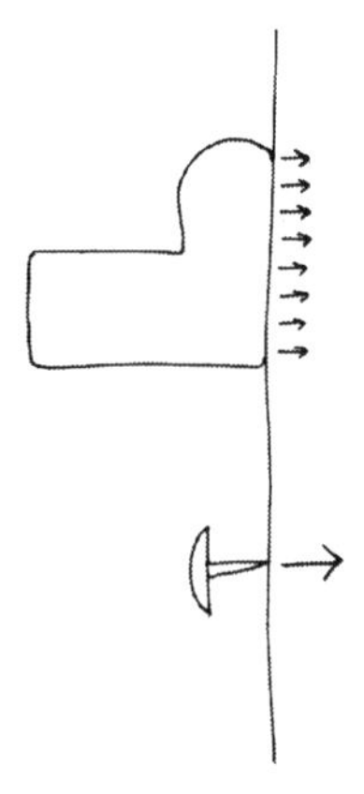

If force is applied over a **smaller area**, the **pressure** is **high**.

Pressure can be calculated with the following formula:

$$\text{PRESSURE} = \frac{\text{FORCE}}{\text{AREA}}$$

Area will be in metres squared (m^2). We already know force is measured in Newtons.

Pressure is measured in **Pascals** (Pa) after the French Physicist, Blaise Pascal.

PRESSURE IN FLUIDS

Have you noticed, when you swim underwater, the **pressure** you feel in your ears? This pressure increases as you dive deeper and it is the reason deep-sea divers need ear-protection.

The deeper an object is submerged under water, the greater the mass of water will be pressing down on it.

Try this (with adult supervision)

Using an empty 2 litre drink bottle, pierce 5 holes down the side, at roughly equal distance apart using a hammer and nail. Now, over a sink or bath, fill the bottle with water. Do not replace the lid.

What do you notice about the angle of water squirting out of the different holes?

What conclusions can you draw about the effect of *depth* on water pressure?

The same principles apply to an object submerged in any **fluid**. A fluid can be a liquid or a gas. A fluid is simply something which is able to change shape and flow to fill a container and assume its shape.

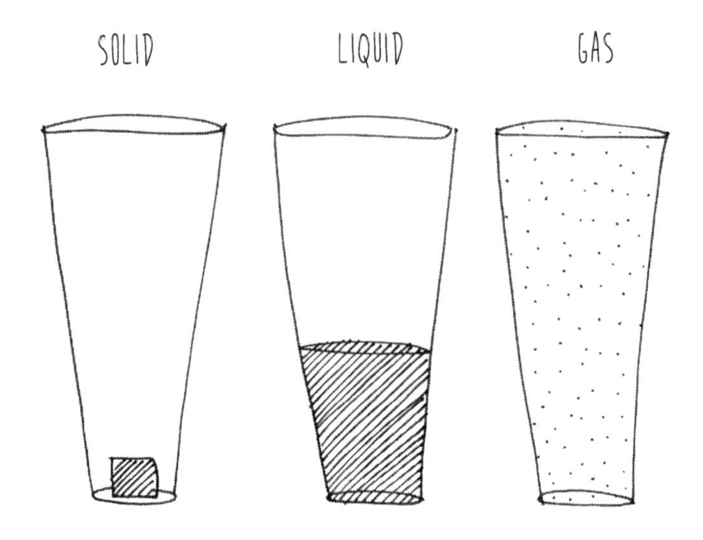

A fluid can be a liquid or a gas

Fluids exert pressure on their surrounding surfaces or boundaries. When a fluid exerts a pressure at 90° we say it is exerting a **normal** pressure.

Liquids exert pressure in all directions, including upwards. This upward pressure is called **upthrust** and is the reason objects float. As an object is placed in liquid, it begins to sink, but the upthrust increases until the object is forced to float.

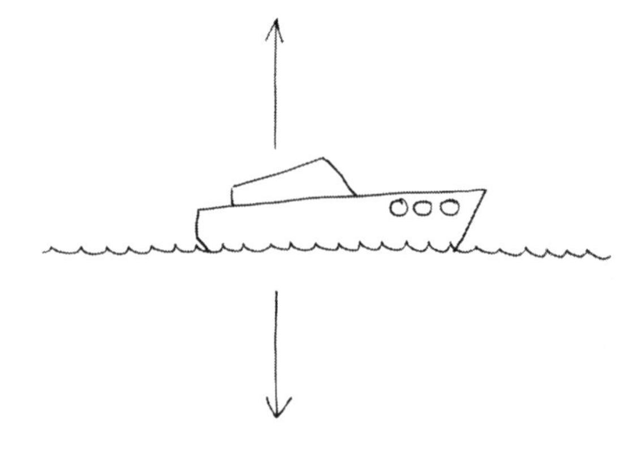

The object's weight and the liquid's upthrust will oppose each other. If the weight is greater than the upthrust then the object will sink. The upthrust on a floating object is equal and opposite to its weight. (Remember balanced forces?)

ATMOSPHERIC PRESSURE

The atmosphere exerts a pressure on you and all the things around you. As with pressure in fluids, this is caused by the weight of atmosphere above us. It is called **atmospheric pressure**. Also, similar to pressure in fluids – it increases and decreases depending on the object's height relative to the weight of air above.

For example, the atmospheric pressure at sea level is approximately 100,000Pa and at an aeroplane's cruising altitude may be only 20,000Pa.

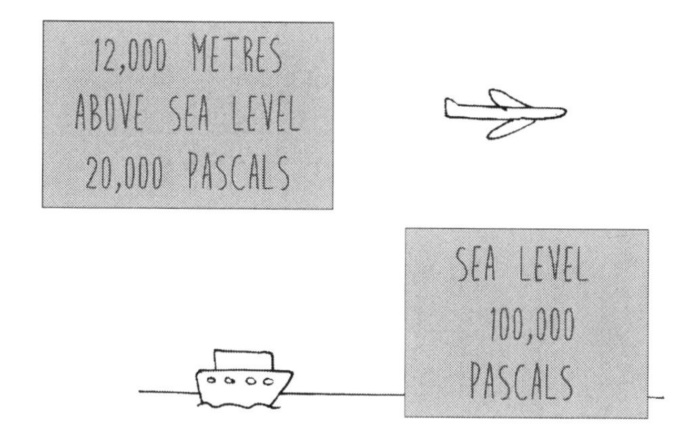

This pressure can be measured with an altimeter which is used in airliners to determine their height. The same technology is used in many step-monitor watches to give a reading of how far you have climbed. However, during storms and great fluctuations in air pressure, the watches can become inaccurate and give very exaggerated readings.

SPEEDY READING

I hope you didn't read too quickly (and miss out the fun stuff), but if you did set a timer at the start of this book, then now's the time to stop it. You can type 4373 into a calculator and divide it by your time in minutes .... and there's your answer in words per minute.

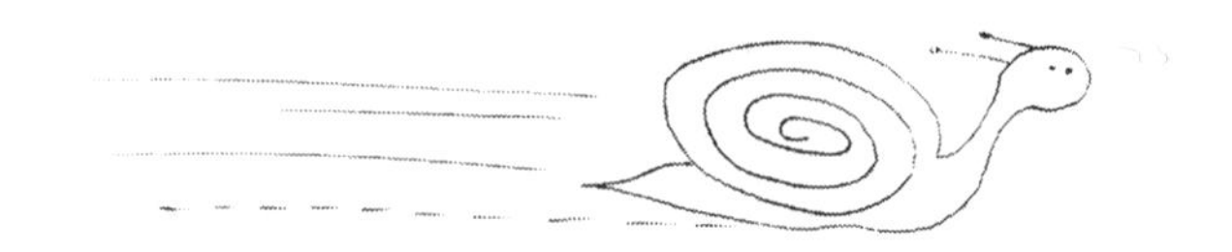

GLOSSARY

Acceleration
The rate at which something changes speed.

Air Resistance
See Drag

Atmospheric Pressure
The pressure exerted by the weight of the atmosphere. At sea level this is approximately 100,000Pa.

Average Speed = distance travelled ÷ time taken

Balanced Forces
When two forces are acting equally but in opposite directions.

Contact Force
A force that acts by direct contact.

Deceleration
The rate at which something changes speed in a negative direction (ie, slows down).

Deformation
A change in the size or shape of an object due to an applied force.

Direction
The course along which an object moves.

Distance
The length of space between two objects (measured in metres).

Drag
Also known as Air Resistance, this is the friction causes when an object moves through air.

Elastic
An object able to return to its usual shape after being squashed or stretched.

Elastic Limit
Maximum point to which an object can be stretched, beyond which it will be permanently deformed and unable to return to its original shape.

Fluid
A substance that has no fixed shape and takes on the shape of the container holding it (gas or liquid).

Force
A push or a pull on an object resulting from the object's interaction with another object. We measure force in Newtons.

Force Diagram
A diagram showing each force with an arrow, indicating the size and direction of the force.

Friction
A force between two surfaces or objects moving over each other.

Frictional Force
See Friction

Gradient
The slope of a line (specifically on a graph).

Gravitational Force/ Gravitational Field Strength
The force of gravity. On Earth this can be approximated to 10N/kg.

Gravity
A force that attracts objects towards a massive object. The higher the mass, the higher the gravity.

Linear Relationship
Direct relationship between two variables so that when plotted on a graph, a straight line is produced.

Mass
The amount of matter something contains. Mass is measured in kg.

Massive
Having a relatively high mass.

Metres per Second
Unit of measurement for speed or velocity, indicating distance divided by time.

Metres per Second per Second
Unit of measurement for acceleration. Velocity divided by time.

Moment
The turning effect of a force around a fixed point or pivot.

Newtons
Units of measurement for force or weight. Named after the English Physicist Sir Isaac Newton.

Non-Contact Force
A force that acts without direct contact (eg, gravity and magnetism).

Normal
Fluid exerts pressure on all surfaces at 90°. We say this is acting *normal* to the surface.

Pascals
Units of measurement for pressure. Named after the French Physicist Blaise Pascal.

Pivot
The point on which a beam turns.

Pressure
The pressure an object exerts on another object. It is calculated by dividing force by area (kg/m^2) and is measured in Pascals.

Pulling
Exerting a force on another object so it moves towards you.

Pushing
Exerting a force on another object so it moves away from you.

Relative Motion
Resulting speed experienced between two objects travelling at different speeds.

Relative Speed
See *Relative Motion*.

Resultant Force
When several forces act together on an object, these forces can be combined giving the *resultant force*. This single force has the same effect as all the individual forces acting together.

Simple Machine
A basic mechanical device that can change the size or direction of a force.

Size
The size of a force is its magnitude measured in Newtons.

Speed
The rate at which something moves. We measure this in metres per second.

Speed-Time Graph
A graph comparing speed (m/s) and time (s). We can use the gradient to determine acceleration (m/s^2).

Tension
The pulling force, for example in a cable or rope.

Time
Time taken for something to occur (eg, to travel a distance) is measured in seconds.

Unbalanced Forces
Forces that cause a change in the motion of an object (as opposed to balanced forces, which would not cause it to move).

Upthrust
The upward force that a fluid exerts on an object floating on it.

Velocity
The rate at which something moves in a particular direction. We measure this in metres per second.

Weight
The force exerted on an object by gravity. Weight, like force, is measured in Newtons.

Work
When force is applied to an object causing it to move over a distance. Work is measured in Nm (Newton metres).

14123212R00025

Printed in Great Britain
by Amazon